HON·SOIT·QUY·MAL·Y·PENSE

Department of the Environment

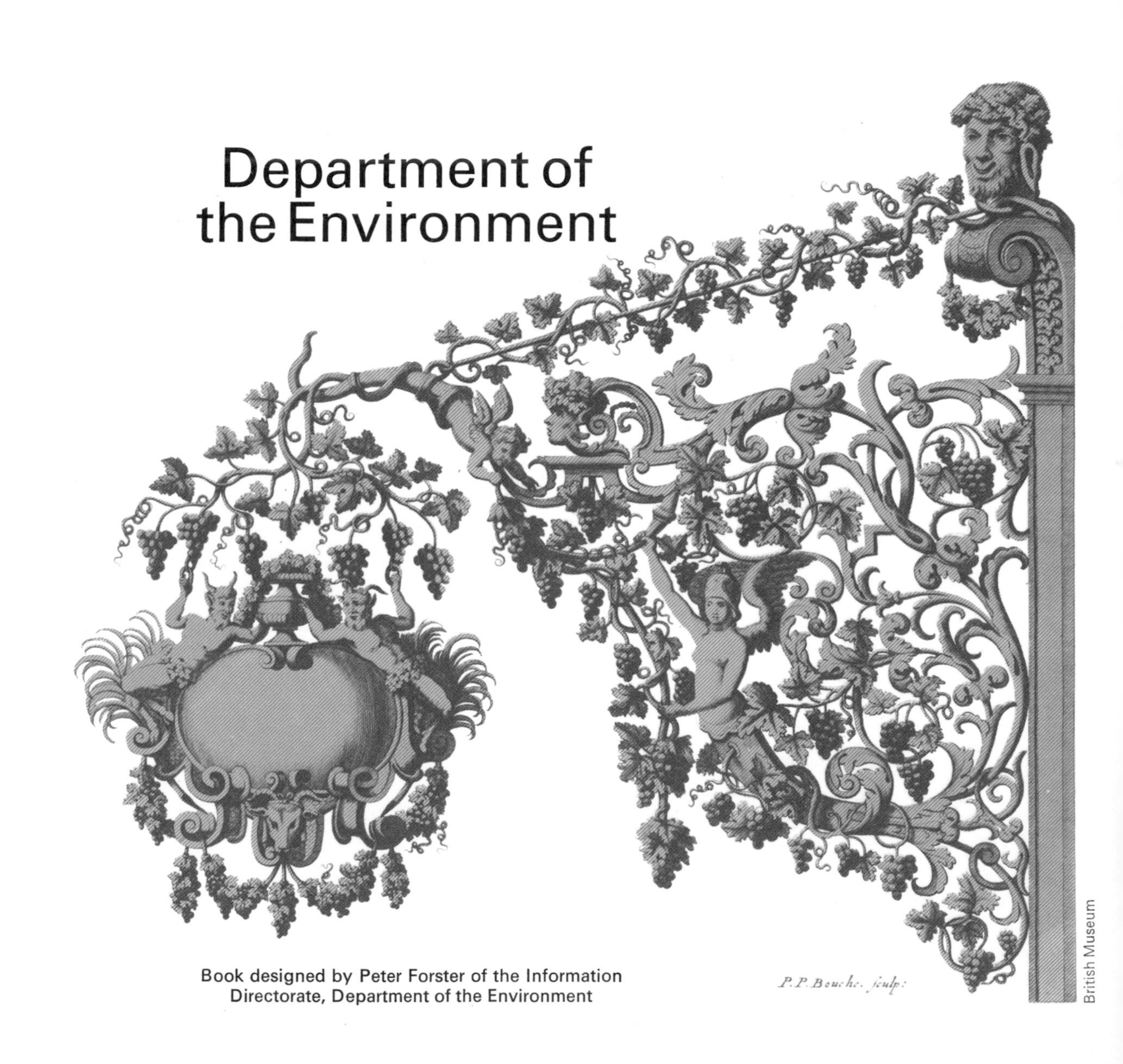

Book designed by Peter Forster of the Information
Directorate, Department of the Environment

THE GARDENS AND PARKS AT HAMPTON COURT AND BUSHY

by
David Green

Her Majesty's
Stationery Office
London 1974

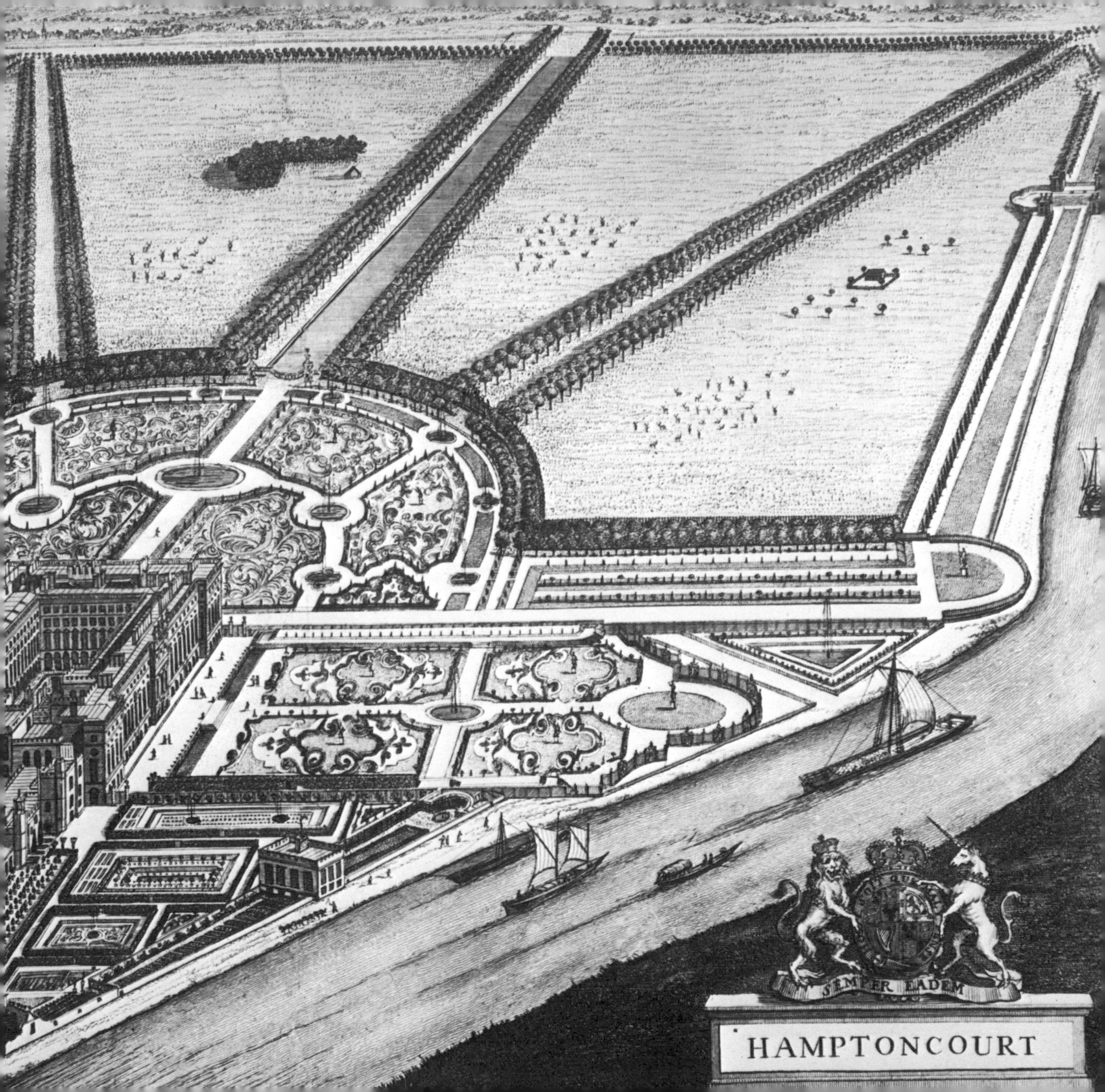
SEMPER EADEM
HAMPTONCOURT

Contents

The design of the cover is based on Sutton Nichols' engraving of the Great Fountain Garden, c. 1695.

The endpapers show four panels of the Tijou Screen.

On the two previous pages a modern aerial view is contrasted with the same easterly view imagined by J. Kip, c. 1700.

The illustrations on this page and on the title page are taken from Tijou's "Nouveau Livre de Dessins" of 1693. Opposite: The south front.

HONI SOIT QUI MAL Y PENSE
DIEU ET MON DROIT
r youth,
wayd:
down truth,
d:
t deed,
reed.
Behoulde the figure of A Royall Kinge,
One whom sweet victory euer did attende:
I From euery parte wher he his power did bringe,
He homewarde brought yͤ Conquest in yͤ end.
And when yͤ fates his vitall thred had spunne:
He gaue his glory to A Vertuous Sunne.
But Sorow care & ciuill broyles lykewise.
This Sacred Queene ELIZABETH exylde:
Fallshood did fall before her Gratious eyes,
3 And perticution turn'd to mercy mylde.
Plenty and peace throughout hir dayes are seene,
And all the world admyr's this mayden Queene.

The Tudors. Henry VIII hands his sword to Edward VI. Beside him stands Mary I. In the foreground Elizabeth, dressed in the fashions of 1595–1600, is attended by Peace and Plenty. A detail from an Allegory of the Protestant Succession engraved by William Rogers.

Part 1 The Past

Close by the meads forever crown'd with flowers,
Where Thames with pride surveys his rising towers,
There stands a structure of majestic frame
Which from the neighb'ring Hampton takes its name . . .

When, in 1525, Henry VIII acquired Hampton Court from Cardinal Wolsey it was said to be 'more like unto a paradise than any earthly habitation'. The air was of an 'extraordinary salubrity', the gravel soil healthy, while the whole domain of buildings, walks and meadows lay invitingly in a crook of the Thames.

The Thames indeed was the genius of the place, the key to that kingdom, a living thread joining Whitehall to Hampton, where it girdled the grounds, filled the fish-ponds and fed the fountains, themselves a delight and a refreshment to the spirit.

Hampton Court was then a place for meditation. Wolsey, pacing the gardens of a summer's evening, read his office; and later, before his fall, he was seen strolling there, the King's arm flung across his shoulders.

Those Tudor gardens were modest enough compared with what was to come. One landed there, on the south-west, at the Water Gallery, to make one's leisurely way, by pond and herb gardens, to the south front. The King soon took a notion to raise a riverside mount, crowned with a gazebo from which to look down on the Thames and the water-meadows or on the palace and its gardens. To reach that summer-house, 'without pain', Henry commanded them to make a spiral way: a 'cockleshell walk', at every few steps flanked with what were called the King's Beasts – heraldic models (lions and leopards, harts and antelopes, dragons and unicorns) all painted and gilded and mounted on posts or pedestals. But wherever you walked in Henry VIII's garden you would meet with these emblems, some grasping vanes embellished with roses or fleurs de lys; and there were sundials too, sixteen of them, their dials embossed with a crown and the royal arms.

In his first year of ownership Henry planted a flower garden, a kitchen garden and two orchards. Queen Anne's Bower, an arched alley of hornbeam on a terrace overlooking what is now the Privy Garden, was named after Anne Boleyn. It was a walk which in time was to change its trees to wych-elms and its name to Queen Mary's Bower, after William III's queen who liked to sit and sew there with her ladies-in-waiting.

We hear nothing of Tudor gardens east or west of the Palace. On the north Henry VIII laid out his tiltyards with five observation-towers (one survives) for watching the jousting. Later these became kitchen

Left: The Diana Fountain. Right: Charles II, engraved by R. White after a design by A. D. Hennis. (The title page of Anthony A. Woods' "Historia Antiquitatis Universitatis Oxoniensis" of 1674.)

gardens and, later still, rose gardens and tennis courts. Henry's covered court for 'real' tennis is still to be seen on the east front and may be entered from the Broad Walk, near the Wilderness Gate.

Queen Elizabeth I – although on her first visit as princess she was virtually imprisoned by her sister, Queen Mary, in the Water Gallery – had a fondness, as Queen, for Hampton Court. Of a winter's morning she was to be seen walking briskly in the grounds 'to catch a heat'; or she might in summer be glimpsed looking down through a leaded window at the knot-garden she had had planted close to the south front. The Elizabethans' knot-gardens – 'knots so enknotted it cannot be expressed' – were often as intricate as their poems and madrigals. The dwarf box borders writhed and intertwined, while the beds themselves were packed with herbs and flowers – lavender and candytuft, pinks and sweet williams – never allowed to grow higher than the box about them. The effect, as one looked down on such a garden, was of cut velvet, the flower-beds set off by their box-edging, by the paths of raked sand which ran between them and by low walls covered with rosemary. There will almost certainly too have been a fountain, for the Queen liked her gardens to have water. She liked also bowers or *berceaux* to sit in – in the sun, out of the wind – when not busy with state affairs or out hunting with the long bow in the wild park. Rose trees were at that time fourpence a hundred and weeders could be hired for threepence or fourpence a day.

Time passes, the leaves fall and, for a hundred years (until the reign of Charles I), change is left mainly to the seasons. Then in 1638 Charles has a tributary of the Colne diverted through Bushy Park to make the Longford River, and at about the same time he encloses, between Hampton and Richmond, ten miles of deer park, and commissions from Francesco Fanelli the Diana Fountain.

Of all the many fountains the courts and gardens of Hampton Court have been enriched with, this of Fanelli's was, as a work of art, the most distinguished: an elaborate conceit of boys with dolphins, and sirens with spouting breasts, the whole edifice crowned with a bronze of Arethusa, mistakenly called Diana. In its original position in the Privy Garden, not far from the Thames, it was appropriate and easy to admire. Between 1689 and 1694 it was much altered, and again, towards the end of Anne's reign, Wren was commanded to have parts of the fountain re-cast and the entire work removed to the Chestnut Avenue pool in Bushy Park. There, on its high plinth enriched with 'frostwork', it remains, too distant to be appreciated.

On the credit side, however, there was and is the Longford River, more than nine

National Portrait Gallery

The east front of the palace – as it was in Tudor times and as it is now.

miles long, which provided a new water-supply not only for Bushy Park, which it enters at Hampton, but for the Palace, its gardens and its fountains.

When in 1653 an inventory of the estate was ordered, with a view to selling the late King's possessions, strange customs came to light, as for example that one William Hogan had been granted by James I 'the office of keeping the two little new gardens at Hampton Court, next adjoining to the Thames side, and the distilling of all herbs, waters etc., together with the distilling house and other houses within the said gardens . . . . and the late bowling-alley adjoining thereto'. Herbs for medicine had been grown there by monks before the coming of Wolsey and, nearly two hundred years later, a fumigation of rosemary, thyme, wormwood, southernwood and other herbs was described for William III's dropsical legs.

In the event the royal estate was retained for Cromwell, who improved the Longford River and, in the eastern part of Bushy Park, made the Hare Warren Ponds (now the Heron and Leg-of-Mutton Ponds). There was no time or inclination for gardening; so that at the restoration of Charles II in 1660 all the royal parks and gardens needed to be remade.

''tis certain that Prince', wrote Stephen Switzer in 1715, 'did plant the large Semicircle before the Palace at Hampton Court in pursuance of some great Design he had form'd in Gard'ning'; and it is equally certain that Charles II was responsible for the Long Water canal and for the three great avenues of lime-trees (from Holland) which radiate eastward from that semi-circle to form a gigantic goose-foot or *patte d'oie*, stretching or seeming to stretch to the horizon.

This was no English pattern. In scale it was more typical of Louis XIV and of his landscape-gardener André Le Nôtre, whom Charles had invited to England in 1660. The reason for Le Nôtre's not coming is still obscure. He is known to have designed a parterre for the Queen's House at Greenwich, but the goose-foot avenues in the park above it spoke of Le Nôtre's former master, André Mollet, who planted for Charles a third goose-foot (the middle toe a canal, as at Hampton Court) in St James's Park. Mollet, recommending in his book *Le Jardin de Plaisir* (1651) this form of goose-foot, suggests that the toe-avenues should radiate from a double semicircle of trees – an obstacle to the prospect which Le Nôtre would never have tolerated; and this, as contemporary plans of Hampton Court, Greenwich and St James's Park show, was in fact precisely the form adopted in all three cases.

In June 1662 John Evelyn, revisiting Hampton Court, found the park formerly a naked piece of ground, now planted with

sweet rows of lime trees, and the canal water now near perfected'. All this was on the east: a great arrow of trees, its point at the very heart of the east front. Strolling to the south, Evelyn admired Fanelli's fountain and 'a parterre which they call Paradise, in which is a pretty banqueting house'. He qualified his admiration however by adding: 'All these gardens might be extremely improved, as being too narrow for such a palace.'

From lessons learned in exile Charles, though his means would not run to a Versailles, adopted for Hampton Court as far as was practicable the setting of a royal *château*: on the south the *jardin d'honneur* and on the east, where the Thames was more distant, subjection of the landscape for as far as the eye could see: impressive, if perhaps, with vast plains between the avenues, a little dull. However, the Home Park was 'filled with all sorts of beasts of the chase' including, of course, deer; and that, for Charles, may have been what mattered. In his brother's reign, short and disastrous as it was, Hampton Court was again neglected. This was more than made up for, however, by James II's successors, William and Mary, who paid the palace more attention than it had had since it was built. As Switzer put it, 'The Foundation of Great Designs being laid at Hampton Court by their Royal Uncle King Charles II, it was thought to be one great Inducement to those

Sir Christopher Wren, engraved, after Sir Godfrey Kneller, by T. Holloway in 1798.

National Portrait Gallery

Princes (William & Mary) to take up their chief Residence there, and Gard'ning soon felt the happy Effect of it. The Great Garden, the Garden next the River, call'd now the Privy Garden; and Wilderness and Kitchen-Gardens (both on the north) were made with great Dispatch. The only fault was the Pleasure-Gardens being stuff'd too thick with Box, a Fashion brought over out of Holland by the Dutch Gardeners, who us'd it to a Fault, especially in England where we abound in so good Grass & Gravel. But upon all other Accounts the Gardens were noble, and their Majesties' Designs yet nobler . . . '

To support them there was no shortage of experts, from Wren and Talman (Comptroller), Daniel Marot (William's architect in Holland) and the Earl of Portland (Superintendent of the Royal Parks and Gardens) to the gardeners George London and Henry Wise, who had succeeded Charles II's men: Mollet, Beaumont and John Rose.

From the outset in 1689, when Wren was called upon to rebuild the east front, William seems to have set his heart upon what came to be known as the Great Fountain Garden. This was Charles II's semicircle, but enriched with beds of box scrollwork 'like lace patterns', probably designed by Marot, and with no fewer than thirteen fountains, all to be seen and heard playing as one sat in one of the state apart-

William III and Mary II, engraved by R. White.

ments then being built. For after all, why should Louis XIV have the monopoly of fountains? As Switzer remarked, 'Considering how beautiful an addition water is to gardening, 'tis hardly to be purchased too dear, being indeed the Life and Spirit of all Country Seats, without which they are dull and flat'.

It was a grandiose work and one which made the prospect from the Palace eastward infinitely more interesting (page 28). To make room for the new garden the Long Water was, as it neared the Palace, curtailed and, as though by way of compensation, given an appendix, called the Lower Wilderness, at its eastern extremity. The beds of *broderie* (dwarf-box arabesques set off by crushed brick or sand) were edged with yews and white hollies, trimmed respectively into obelisks and globes.

To match this magnificence, Jean Tijou (spelt 'Tissue' in the accounts) was commissioned to make twelve wrought-iron panels, each ten and a half feet high and over thirteen feet broad, as a screen to stand in the gardens. Judging from early engravings, this screen was made for the site it now occupies, at the southern end of the Privy Garden, next to the Barge Walk beside the Thames. Some authorities maintain, however, that at one time they flanked the Tijou gates on the east, at the entrances to the lime avenues. In the nineteenth century they were sent to the Victoria and

National Portrait Gallery

A detail of the Tijou Screen and an engraving of gate and screen from Tijou's "Nouveau Livre de Dessins".

Albert Museum, where they remained until re-erected in their present position in 1902. Tijou published his designs for two of the panels in his *Nouveau Livre de Dessins* of 1693. Nothing more princely nor more beautiful has ever been made in iron. Tijou's bill came to two thousand one hundred and sixty pounds two shillings and a farthing.

Bearing in mind William's frequent absences in Holland (he spent in fact only one summer in England throughout his reign), the amount of time, thought, labour and money expended in the remaking of Hampton Court and its grounds was astonishing. 'Yet in the least interval of Ease', we are told, 'Gard'ning took up a great part of his Time, in which he was not only a Delighter but likewise a great Judge.' His queen, too, 'lost no Time but was either measuring, directing or ordering her Buildings; but in Gard'ning, especially Exoticks, she was particularly skill'd, and allowed Dr Plunket £200 per annum for his Assistance therein'.

In her zeal for exotics Queen Mary sent plant-hunters to Virginia and to the Canaries. Naturally in that age of symbolism, she, married into the House of Orange, had an orangery (on the south front) full of orange-trees. She went further and commissioned one Josias Iback to make her 'the resemblance of an orange-tree cast in brass'. Sculptors of standing – Cibber and Pearce – strove to excel each other in the carving of giant urns and fountains.

While the east front was rebuilding, Mary happily set up house in the Water Gallery, with its balcony above the Thames, near the southern end of the Privy Garden, then remodelled as a terrace-bordered parterre. This gallery charmed her. Its times of melancholy – when Elizabeth was a prisoner there or when Charles I was escaping from Cromwell – were forgotten and now, as she planned it, it was to be 'the pleasantest little thing within doors that could possibly be made'. There she would have her Looking-Glass Closet painted with animals. There would be Marble and Bathing Closets, a dairy and the Queen's Delft Ware Closet, in which many of her favourite pieces would be displayed on overmantels designed and carved by Grinling Gibbons.

On a summer's day Queen Mary could sit and sew in the bower on the western terrace of the Privy Garden; or if plants had arrived from abroad she could see them safely stowed in the Glass Case (or Glass Frame) Garden, which ran the length of the eastern terrace and had a small tower let into the high wall. But her time was pitifully short. When in 1694 she died of smallpox William was too dispirited to continue with his building and gardening projects. Only the fire which four years later destroyed Whitehall Palace could force him to

Henry Wise, the painting by Sir Godfrey Kneller at Kew Palace. Far right: Plan of the Wilderness and, above, the Wilderness, from Knyff's View (see centre pages).

review his plans for Hampton Court. Then indeed his enthusiasm was rekindled; so much so that, had he lived longer, the whole of the rest of the Palace would probably have been rebuilt. Grand as Wren's design had been for the east front, his design for the north was even more ambitious, entailing as it did a noble, colonnaded entrance and, leading to it, a mile-long avenue of horse-chestnut and lime.

In September 1699, Talman, William's Comptroller writes to him: 'We are making a road 60 feet broad through the Middle Park and a Bason [for the Diana Fountain] of 400 foot diameter in the middle of the circle of trees, which will be very noble. We have abundance of our projects (if His Majesty will like them) by several noble lords that we here call the Critiques. . . .' Wren might have preferred to deal with the King alone, but the King was in Holland and one of his Dutch favourites, William Bentinck, Earl of Portland, was Superintendent of the Royal Parks and Gardens. However, time and money ran out; so that at William's death in 1702 the north front of the Palace remained unaltered, although the state approach to it – Wren's Chestnut Avenue through Bushy Park from Teddington, and the pool for the Diana Fountain – had by then been made.

Between the Chestnut Avenue and the front it was aimed at lay the Wilderness, a large formal garden of clipped evergreens, planted for William by George London and Henry Wise. Except that it included a circular labyrinth called a troy town, it was anything but what we understand by a wilderness today. The yews, the box, the hollies were planted to form a geometrical pattern. In Wren's first plan no maze was included and, to judge from Henry Wise's bills, it was not until the last year of Anne's reign (1714) that the famous triangular maze, more or less as we know it – 'a figure hedge work of large evergreen plants' – was planted near the Lion Gate, where it stands today. Switzer criticised the maze as too simple and easy. It had only four stops. He could have provided one that had twenty. Others condemned the 'regular straight walks' of the Wilderness in which, they declared, one might 'faint for shade in a sultry day'. There were plenty, however, to admire and they included Defoe. 'Nothing of that kind', he wrote, 'can be more beautiful.'

All this was on the north; but on the south, too, great changes were made in the last two years of William's reign. Henry VIII's mount (10 000 cubic yards of earth) was levelled and then, by royal command from Holland in 1700, the Water Gallery was demolished. This indeed was a loss, though one compensated for in part by the building of a small, castellated banqueting-house, which still stands there beside the Thames, a lodge enriched with magnificent

mouldings by Grinling Gibbons, some of them salvaged from the Water Gallery when it was pulled down. In the following June Henry Wise was ordered to lower the Privy Garden by ten feet, to afford King William a better view of the Thames. While that garden was being remade, the trees and shrubs (hornbeam, cypress and so on) growing in it were transplanted to the Wilderness, with the intention of moving them back again when the Privy Garden was ready. Wise was skilled in transplanting. Even so it is hard to imagine hornbeam taking kindly to it; and that may be why, on the terrace called Queen Mary's Bower, one sees wych-elms instead of hornbeam today.

Among the last of King William's schemes was the Great Terrace beside the Thames, a straight half-mile walk from the southern end of the Broad Walk to Wren's quartet of rectangular pavilions set about a lawn for bowls. Switzer pronounced it 'the noblest work of that kind in Europe'. Yet it was modest enough compared with what William had planned to build for his queen, had she lived. Talman's plans for an elaborate 'Trianon' at Ditton may still be seen in the R.I.B.A. Library. But that was not to be; nor, thanks to a molehill (if that tale be true), was William ever to fulfil his ambition of making Hampton Court the chief residence for a line of kings.

In 1702, on the accession of Anne, several factors together inclined her towards a modification of her predecessor's ambitious designs. She had always disliked her brother-in-law, William; and his Dutch favourites, including the Earl of Portland, were soon dismissed. It happened, too, that she could not abide the smell of box. In July 1702 she took a resolution 'to restrain the expense of the Gardens'. As time went on she did in fact spend a great deal on gardens, but for the most part it was Kensington and Windsor that benefited rather than Hampton Court, which she used mainly for Cabinet meetings.

Of the Great Fountain Garden on the east Defoe wrote, in 1724: 'This part of the garden was at first laid out in a parterre of scroll-work in box, which was not only very costly at first making but was also very expensive in keeping constantly clipped; which, together with the ill scent which frequently reached to the Royal Apartments, occasioned its being demolished and the ground disposed into another form. And if at the same time', he adds, 'all the shorn evergreen trees had been thrown out and a finer disposition made of the ground, it would have much better corresponded with the noble apartments which overlook it than it does at present.'

But from the plans which survive it seems clear that William's garden was altered and modified, rather than demolished. The

The Royal Palace of Hampton Court, an 18th-century view. Over the page: Hampton Court in the reign of George I, a detail from a painting by Leonard Knyff. By gracious permission of H.M. The Queen.

outer semicircle of eight fountains vanished, leaving a continuous semicircle of water called the Little Canal. On the far (eastern) bank of that canal stood 403 lime trees, some of them four and a half feet round and all planted in the reign of Charles II. In 1703 Henry Wise was ordered to transplant all those limes to the western bank. With much labour and skill he did so. The purpose of the operation, however, remains far from clear. Why should Queen Anne wish the trees to mask the canal and to enclose her garden? Could the reason have been, at least partly, psychological? She cared for nothing that might put her in mind of William. Furthermore, she preferred small houses and small rooms to large ones. At Kensington and at St James's she lived in small closets, while at Windsor she much preferred her Little House (sometimes called the Garden House) to the Castle. At Kensington Palace she had Wise make her a sunk garden out of a gravel-pit. Her taste in landscape was the very reverse of Louis XIV's.

In the meantime the beds of scrollwork had made way for grass and gravel, enlivened with only five fountains and four statues (of Apollo, Diana, Saturn, and the Gladiator – the last from St James's Park) on grass plots bordered with trimmed yews and hollies. These last, it has been suggested, formed the initials of William and Mary; but had that been so, or had Anne known it, they would almost certainly have been rooted out. By the middle of Anne's reign, for one reason or another, William's Great Fountain Garden had become what the French called a *parterre anglais* – a green and pleasant compromise, economical and unexciting – and in a later reign (George II's) the five fountains (we were never much good at fountains) would be reduced to only one.

It was Defoe who remarked that Hampton Court was loved by alternate monarchs; and perhaps it was as well that Anne changed it as little as she did. To Wolsey's buildings on the south-west, near the Pond Gardens, she added a very plain orangery; but the Privy Garden remained much as William had left it, until 1713 when, again for no given reason, Fanelli's fountain was removed to Bushy.

In the very wet winter of 1710–11 Wise was ordered to widen and lengthen the Little Canal. He made it forty instead of thirty feet across and gave it 'arms', northward and southward, though these were still masked by trees. Clipped hollies and yews bordered the Broad Walk (on its western side), as they still did the grass plots of the Great Fountain Garden. In the same winter, too, Wise was commanded to prepare some twenty miles of 'chaise ridings, fit for Her Majesty's passage with more ease and safety in her chaise or

"An oblique View of the East front of Hampton Court with part of the Garden", drawn by Highmore and engraved by J. Tinney. Below: The east front in the mid-19th century, drawn by T. Allom and engraved by T. A. Prior.

coach in both her parks'. In a long, winding course, 'pick'd out of ye most pleasant parts', among limes and chestnuts, mole-hills were levelled and hayseed sown so that Queen Anne, though she had grown hugely corpulent, could career along them in her high-wheeled gig, herself in black cloak and hood. For Wise it was but one more responsibility; for here, as at Blenheim, he was not one to be daunted by heavy commissions. Explaining why he had exceeded one of his estimates he casually mentioned 'a great hill in Kingston Avenue, which much obstructed the view from the house and gardens and was thought proper to be levelled'.

Almost the last work Wise did for Queen Anne at Hampton Court was the Maze near the Lion Gate, a gate strangely typical of her successor George I, whose cypher it bears; the same of whom Lord Chesterfield said that England was too big for him. The stone piers for the gates, bearing Anne's monogram and topped with grotesque stone lions (their counterparts are at Blenheim), have been attributed to Vanbrugh and certainly they are on the Vanbrugh scale; but the gates they carry, though beautiful, are, as Defoe said, pitiful low ones. Tijou made them for another site in the Home Park.

Luckily for Hampton Court, and for us, the Hanoverians, George I and George II, decided to enjoy Hampton Court without spoiling it. From contemporary accounts we catch glimpses of them on the royal barge hung with crimson silk, pursued by their orchestra; or again listening to music at the Bowling Green, where Wren's pavilions had been turned into boudoirs and drawing rooms. In George II's reign Lord Halifax, following in the hazardous footsteps of Charles I and Cromwell, attempted to annexe Bushy Park as his private estate, but was thwarted by Timothy Bennett, a seventy-five-year-old shoemaker of Hampton Wick.

Since the death of George II in 1760 Hampton Court has never been lived in by a reigning monarch; and when George III asked Capability Brown (Surveyor to His Majesty's Gardens and Waters at Hampton Court) to improve the grounds, he is said to have excused himself 'out of respect to himself and his profession'. With Brown's office (he was appointed in 1764, the year he planned the lake at Blenheim) went Wilderness House, just west of Lion Gate, and a salary of £2000 a year. At Hampton Court two small but important works are attributed to him: (i) in the Privy Garden terrace-steps were superseded by grass banks because, as he is reported to have said 'We ought not to go up and down stairs in the open air'; (ii) the planting, near the Pond Gardens, of the Great Vine in 1768. This was from a cutting of the Black Hamburg vine then growing at Valentines

View to the east across the Great Fountain Garden, from the roof of the palace.

near Wanstead in Essex. The Great Vine at Hampton Court has flourished ever since.

William IV, while Duke of Clarence, became Ranger of Bushy Park, lived in Bushy House (later the residence of Queen Adelaide) and interested himself in the breeding of George IV's racehorses. In the Home Park (the Stud House and its paddocks lie north of Long Water) the cream-coloured horses which drew Queen Victoria's coaches continued to be bred from stock brought over from Hanover by George I.

George IV made only slight alterations, removing four statues that graced the centrepiece of the south front and, even more to be regretted, the giant urns sculpted in marble for the gardens by Edward Pearce and Caius Cibber, in the reign of William III. They are still at Windsor. It will be a great day for Hampton Court when they are returned.

In 1838 Queen Victoria opened Hampton Court and Bushy to the public, and hundreds of thousands of visitors began to pour through. In Bushy Park Chestnut Sunday – in the middle of May when the avenue is at its best – soon became a national institution.

Ernest Law, who lived at the Bowling Green House and, in the reign of Victoria, published *The History of Hampton Court Palace,* while praising the gardens in 1890,

hinted that they had to some extent suffered in the pervading blight of carpet-bedding. In the Great Fountain Garden he was sorry to see 'attempts to follow the fluctuating follies of successive fashions in gardening . . . by efforts to vie with the costly pretentiousness of the modern style'. This, he adds, was not only extravagant but incongruous at Hampton Court: where 'the more the original arrangement is preserved . . . the greater is the benefit to the character of the place'.

Luckily, when it came to planning the long border beside the Broad Walk, the influence of William Robinson prevailed and the grouping of plants with respect to height and colour was given serious thought; so much so that in Law's *Hampton Court Gardens* (1926) he gave unstinted praise to the complex planting, at the same time mentioning every plant by name.

What Law liked especially at Hampton Court was a sense of permanence in change, the thought that 'as around us so much still endures unchanged, all things that have been and will be are indissolubly linked with what succeeds and that time itself is but the ever-varying aspect of eternal things'. For good reasons we long to preserve these stepping-stones which link us with what Churchill called 'the far backward of time'. With buildings this need not be difficult; but with parks and gardens there are problems, notably the growth and decay of trees. Queen Anne masks Wolsey's walls with an orangery. They are still there. But what of William's Privy Garden, its parterre obliterated by shrubs and yew trees; or on the east the gigantic goose-foot, imprinted by Charles II, now bosomed so high in trees as to be all but unrecognisable? In the Great Fountain Garden one fountain serves for the thirteen William intended for it; while the yews planted as small obelisks there have, as Wise said of his fruit trees, 'shot to admiration' – and they are admirable, even though bulky and out of scale.

In time, when trees die as they must and are replaced, the problem may tend to resolve itself. In the meantime, to compare Kip's (page 5) with a modern bird's-eye view (page 4) is to see the extent to which nature has taken over. On south and east the patterns laid down by the Stuarts can still, though dimly, be traced. On the north-east William's Wilderness, except for the Maze, has vanished: nature has been given its head. As a demonstration of British compromise – the French-poodle parterre transmuted into an old English sheepdog – there could be no better example; and truth to tell, few visitors today would prefer it otherwise. Parterres were made to be viewed from upper windows. At ground level, as we wander in the Wilderness or in the lime avenues of the Home Park, we tend to forget the original pattern and its purpose. Nevertheless it is important that the outline at least of that historic imprint should be preserved. Great gardens were never made without thought and labour. They deserve to be appreciated and understood.

Part 2 The Present

The best guide to a garden is the man or woman who made it. At Versailles Louis XIV took the precaution of writing his own guidebook: *Manière de Montrer les Jardins de Versailles*. He was determined that no one should view those great gardens without contemplating every wonder in them: each statue and urn, each grove, each fountain; and to that end, in his guidebook he every so often prescribes a pause . . . And so it should be at Hampton Court – a place for strollers and sitters, not for people in a hurry.

No monarch has left a record of how best to see these gardens. Within the palace we may meet the ghost of Catherine Howard. In the grounds we are more likely to sense the shadow of the 'royal' gardener, Henry Wise, in some ghost of a terrace or faint outline of a forgotten parterre.

Leaving the palace by the middle doors of the east front we find ourselves on the Broad Walk in the Great Fountain Garden where, from the reign of Charles II, armies of gardeners have worked to make and keep this garden worthy of the royal apartments that look out on it. There is one central fountain; and the yews planted as small obelisks have grown into cone-shaped trees – black pyramids, Virginia Woolf called them – some thirty feet high. Every summer they are trimmed by hand with secateurs. In good weather two men with a ladder can trim one and a half trees in a day. These yews mark the beginning of the giant goosefoot pattern, continued on the far side of the Little (semicircular) Canal in the form of lime avenues, the middle toe of the goosefoot being the canal made by Charles II and called Long Water. The wrought-iron gates at the canal-side entrances to these lime avenues are by Jean Tijou. His magnificent screen is at the southern end of the Privy Garden.

To reach the Privy Garden from the east front turn right on leaving the palace and then, after 50 yards, right again through the gateway in the wall. The Privy Garden is then on your left. This is the south front of the palace, where William III built an orangery. The sundial on the terrace bears the cypher of William and Mary. The Privy Garden when first made was a formal parterre of flowerbeds centring upon the Diana Fountain. Now, yew and magnolia, lilac and judas-tree have been given their freedom. The high terrace (at eye level before you climb the steps) on the west of this garden is called Queen Mary's Bower. Eleven shallow steps lead to a path shaded by pollarded wych-elms, some of them hollow, the sunshine filtering through and between the leaves. Looking towards the narrow opening at the far end you may glimpse a flight of gulls, for the Thames is just beyond, with the Barge Walk and Tijou's Screen. The twelve huge, wrought-iron panels are in splendid condition. In the reign of Victoria they spent some years in the Victoria and Albert Museum. One makes out harps and thistles, masks, scrolls, roses, a Garter badge . . . and at the same time looks through the screen at the river and the trees beyond: silvery willows and green poplars set off by blue sky.

If you go back along Queen Mary's Bower and then turn left you will come to the Tudor Gardens: the Pond Gardens on the left, the Knot Garden and Queen Anne's Orangery on the right; the path eventually leading to the Vinery and the Banqueting House. From spring to autumn the Pond Gardens make one of the most colourful places in the whole of the grounds. The larger of them, seen through a screen of lime and wistaria, centres upon a curious, small fountain with water playing over a great tussock of long grass. On a hot day in August one feels cooler just for looking at it. There are flagged paths across short turf and, grouped about the fountain, box bushes once trimmed into birds now look fashionably 'abstract'. All about this group and at various levels grow the brightest of flowers: in spring, tulips and hyacinths, jonquils and wallflowers; in summer, stocks and zinnias, geraniums and begonias, heliotrope and dahlias, African marigolds and ageratum. 'Oh look at this, come and look at this garden', says a voice with an American accent. 'This is quite some-

N
Canal Plantation
Pantile Bridge
Bushy House
Park Road
Sandy Lane
Longford River
High Street
CP
Cobbler's Walk
Cobbler's Walk
Warren Plantation
Duke's Head Passage
Broom Clumps
Half Moon Plantation
Leg of Mutton Pond
Church Grove
Chestnut Avenue
Triss's Pond
Heron Pond
Bushy Park
Bushy Park
Kingston Bridge
Barge Walk
Hampton Court Road
Lime Avenue
Hampton Wick Pond
LF&M
Tagg's Island
Coach Park (summer only)
Hampton Court Road
Ash Island
The Green
LM
LF
Wilderness
LF&M
LF
LM
CP
Hampton Court Park
Hampton Court Bridge
The Long Water
CP Car Park
LF Lavatory (Women)
LM Lavatory (Men)
The Diana Fountain
The Flower Pot Gates
The Lion Gates
The Maze
The Trophy Gates
The Tilt Yard Gardens
Cafeteria, Kiosk and Restaurant
The Great Vine
The Banqueting House
The Pond Garden
The Privy Garden and the Tijou Screen
The Great Fountain
Hampton Court Station
The Rick Pond
Raven's Ait
Barge Walk
Rabbit Warren
River Thames
Thames Ditton Island
Boyles Farm Island

thing.' It certainly is. The lead statue of Venus, in the yew arbour at the far end of this garden, was found at Windsor in the nineteenth century in a corner of Mrs Grundy's Gallery, its face to the wall. The smaller of the Pond Gardens, next door, is no less remarkable. This too is a sunk garden, remade within low walls dating from the sixteenth century. More formal than its neighbour, it has a single-jet fountain in the midst of a seventeenth-century pattern of fleurs-de-lys. Notice the way this miniature parterre has its lower walls lined with clipped yew. They are part of the background yet their texture and colour add a great deal. The screen here through which one looks down into the garden is of pollarded hornbeam; and beyond the end of the garden is the castellated roof of William III's Banqueting House. The interior, with its Verrio frescoes and Grinling Gibbons mouldings, is well worth seeing.

On the other side of the path, opposite the Pond Gardens, are the Knot Garden and Queen Anne's Orangery. Ernest Law, the historian who designed this Knot Garden and planted it in 1924 wrote: 'The patterns of the interlacing bands of ribbons are taken entirely from those designed and published by the old masters on gardening of the time of Elizabeth and James I. The interspaces are planted with such old English flowers as tulips, hyacinths and daffodils.' Dwarf box is of course used freely, as are also thyme, lavender and cotton-lavender. In the summer the divisions are filled with low-growing exotics such as ageratum and small pink begonias which, though not known to the Tudors, nevertheless make good effect in this tapestry of flowers.

Queen Anne's Orangery, like the Banqueting House, is of warm red brick with long white sash-windows, in the upper corners of which house-martins trustfully build their nests. In the corner next to the Knot Garden a magnolia, flowering in August, has grown to a great height. Beside the path there are standard roses and, tucked into the near wall, Pearce's Dolphin Fountain, its clear water trickling into a scallop-shell and basin below. On a summer's day the fine turf before the Orangery is bright green, the sky bright blue behind the carved-brick chimney-stacks and the lead lantern crowning an octagonal Tudor turret.

In the Vinery at the western end of the path grows the Great Vine, the most famous vine in the world. At its base its girth is seventy-eight inches, its longest 'branch' measuring a hundred and fourteen feet. Its pale brown sinews snake round the walls and, as though checked in full flow, form an arch overhead: strange that something so alive and purposeful can be so still. The base is gnarled like briar and yet even from that, green shoots appear. There is a

The Great Vine.

tradition that the deepest roots reach to the bed of the Thames. The Vinery, recently rebuilt, is now supervised by a lady-vine-keeper. The grapes (Black Hamburg) are usually sold to the public at the end of August or the beginning of September, the crop often running to five or six hundred bunches, weighing on an average twelve ounces each.

After seeing the Vinery and the Banqueting House the visitor, rejoining the Broad Walk outside the Privy Garden, may turn right for the Thames-side terrace which once led to the Bowling Green; or he may prefer to turn left, for the Tennis Court, the Wilderness and the Maze. If he makes for the river, across the grass, he will soon find himself in an avenue of old lime-trees, growing from a low terrace beside the Little Canal. This terrace is now no more than a gentle slope and yet in its way it is as evocative as the smell of lime-flowers or of heliotrope on the breeze. The Thames is not yet visible, nor even perhaps is its presence suspected until suddenly, as one looks ahead, a sail seems silently to cross the garden, for all the world as though Queen Mary's shallop (her cypher carved by Gibbons on the prow) had sailed out of its museum at Greenwich to fetch her from the Water Gallery again. The mown way to the left runs between river and golf-course. It is bordered by trees (one sees pied wagtails here and spotted

British Tourist Authority

The Dolphin Fountain.

flycatchers), and there are riverside seats for resting and watching the boats go by – the *Eclipse* (full of children), the *Amble* (for two) or a majestic pleasure-boat aptly called *Her Majesty's*, full of happy-looking people. There are swans and gulls and alders and willows and always everywhere boys fishing. This is a quiet place. Since no radios are played in the gardens, the loudest noise one is likely to hear is the drone of a plane or the cackle of magpies.

Returning by the Broad Walk, visitors who are gardeners may feel envious of the perfect site for a long (in this case a very long) border: a really noble extent of warm brick wall. Here at all seasons full advantage is taken of it, the idea being, as the Superintendent puts it, to ensure a constant succession of colour, from the first winter aconite to the last michaelmas daisy; and indeed it is never more beautiful than in early autumn with the michaelmas daisies (so many varieties) at their best.

The Broad Walk, nearly half-a-mile long, runs from the Thames northward to the Flower Pot Gate (its boys with flower-baskets designed by Nost), beside the Kingston Road. But before reaching that gate, and just after passing the enclosed Tennis Court, one can slip through on the left to the Wilderness, and so to the Maze and the Lion Gate, or westward to Henry VIII's Tiltyard, where a restaurant has been added to what was originally an observation-tower for the jousts. In summer meals may be enjoyed out of doors. Close by grow fir and prunus and tulip trees; and among them stands a graceful sundial from Garrick's villa. It bears the inscription:

DAVID AND EVA GARRICK 1765
HAMPTON-ON-THAMES.
UT UMBRA SIC VITA.

On the dial, amidst longitudes and latitudes, one makes out the words:
WATCH FASTER . . . WATCH SLOWER . . .
But here somehow that hardly seems to matter.

The seven-acre Tiltyard has seen many changes: at one time a kitchen-garden and later a nursery-garden, it is now devoted mainly to the growing of roses. The old-fashioned kinds – Albertine, Zéphirine Drouhin and the rest – respond to the protection and warmth of the old brick walls, honeycombed as they are with nail-holes made by generations of gardeners. Even in August the roses still scent the air as one wanders round, admiring late blooms and reading the labels: Blue Moon, Paul's Lemon Pillar, Chaplin's Scarlet Climber, Madam Herriot, Madam Butterfly, Rosa Banksiensis . . . In the walled garden beyond (through a door in the wall) phlox and hellenium, achillea and michaelmas daisy, zinnia and golden rod are massed about close-cut turf. It looks restful and it is restful because here, as everywhere else in the grounds, there are plenty of comfortable places to sit and admire things; while for the young and indefatigable there are hard tennis-courts and a putting-green.

If there are children in the party they must of course see the Maze – and what could be better? Safe and secret, a labyrinth where they can safely lose their way, knowing perfectly well they could never really be lost. The attendant-in-charge keeps on eye on things all the time. In Stuart days these labyrinths were made to amuse grown-ups; and today too parents, uncles and aunts seem glad of an excuse to solve the puzzle for themselves. The hedges, originally hornbeam, are now mostly yew and privet, more than six feet high. They take up little space but make quite a lot of walking. With luck (or is it skill?) one reaches the middle, with its seats beneath chestnuts, in five minutes. The return journey may take longer. 'We're on the right track now,' says someone. 'No we're not, we're back in the middle. Now how do we get out?' 'Swim.' Grown-ups look as though they're trying not to laugh; children as though being lost for a while could be exciting. 'We've been here for hours', says one cheerfully, 'trying to get out'. The knowing ones, those who have been before, emerge with enough energy left for the swings in the Bushy Park playground, just over the way. Others clamour to be taken to the Woodland Gardens.

The Maze and, opposite, the Laburnum Walk.

After crossing the Kingston Road and entering Bushy Park you can take the footpath on your left to the Woodland Gardens, and the walk – half a mile among old oaks and bracken – is a pleasant one; or if you go by car, you can park on the right of the avenue, beyond the Diana Fountain. Then, to reach the Woodland Gardens you cross the Chestnut Avenue and take the signposted path.

The famous avenue, planned by Sir Christopher Wren for William III and nearly a mile in length, has two hundred and seventy-four horse-chestnut trees, planted forty-two feet apart. They are at their best in mid-May (Chestnut Sunday). Behind the chestnuts, lime trees are ranged four deep in subsidiary avenues. In Bushy Park, which runs to 1100 acres, there are ten miles of lime trees alone, and by no means all of them are old or even mature trees. Thanks to the Advisory Committee on Forestry and to the Superintendent and his staff, one sees in various parts of this park recently planted avenues of healthy young limes (west of the Diana Fountain and each side of the Longford River) and chestnuts (at Hampton Hill), carefully spaced and protected from the deer. Fish ponds formed by Oliver Cromwell are on the east side of the Chestnut Avenue. The Pheasantry and Waterhouse Plantation are on the west, and the Woodland Gardens, begun in 1949, provide a delightful walk through both.

The walks in these Woodland Gardens, casual as they seem, have been thoughtfully planned. Undergrowth has been cleared and grass sown; oaks have been spared and shrubs cleverly positioned. One has the feeling that just beyond that clump of azaleas there may be a pleasant surprise – and there often is. Birds are plentiful – woodpeckers, nuthatches, treecreepers, and in spring and summer blackcaps, chiffchaffs and willow-warblers and a great many more. You may be lucky enough to see a kingfisher, for water (the Longford River) is everywhere. Again it looks natural, and again it has been channelled here and diverted there, to water a plantation and to spring surprises. You may enter the wood by a path that winds among high rhododendrons, opening into glades of tall, straight oaks. You wander on until suddenly, beyond a bank of heather, there is the stream, with beside it an immense poplar and at the foot of the poplar a clump of orange-yellow flowers (*ligularia Clivorum*) like spikes of ragged daisies, full of bees. Somewhere beyond the stream a child is playing *London's Burning*, on a recorder. Poor old London, it seems a very long way off. Across the bridge there's a seat beneath a beech tree and another beneath an oak . . . 'It's a lovely place, this,' says someone; and when you are there you know it: at this very moment this is the place to be.

Certainly the walk through the Waterhouse Plantation can be idyllic; and the Pheasantry too has its hidden delights. There at the stream side grow two swamp cypresses (*taxodium distichum*) and on the bank itself they have grown 'lungs' (pneumataphors) looking like a colony of knobbly stalagmites, some of them two feet high. And there are other vast trees there – plane, ash, tulip-tree – some of them probably planted by Capability Brown. There is too that graceful curiosity (*Metasequoia glyptos stroboides*), a fir thought to be extinct until rediscovered in China in 1947. There are islands with alders and willows. There are water-lilies and dragonflies, ducks and moorhens and Canada geese. There are paths far more beguiling than those in the Maze; and the stream too leads one on until it dives and vanishes between trees. For this is no ordinary stream. The Longford River does in fact dip underground before supplying the Diana Fountain: and again it vanishes beneath the Kingston Road before feeding Long Water in the Home Park. In the woods, in May and June, it reflects azaleas. Later it is overhung with full-leafed trees and ferns. On a fine day one could enjoy wandering in the Woodland Gardens forever. WATCH FASTER . . . WATCH SLOWER – What could it matter? Yet at last one turns homeward, with a glance at the deer (heads just visible above the high bracken) and another at

One of Tijou's gates in the park.

the Diana Fountain, its boys-on-dolphins matched by those with fishing-rods at the pool's edge. Five little girls in velvet caps trot by on ponies. Others play on a maypole-swing just over the way. A big yellow ball (somebody lost it) bobs among the duckweed .... We have to go.

Pigeons perch on the lions of Lion Gate and just beyond (if it's in May) the laburnum pergola makes an arcade of gold. The Maze is still popular (it always is) and outside it a gardener has left a wheelbarrow with his birch-broom in it. On the side of the green barrow, in bold white letters, is painted the word WILDERNESS: the perfect endpiece for the day.

And so we drift back by way of Wilderness and Tiltyard to the Trophy Gate and the Bridge and from the far bank take a last look back towards the roofscape of the Palace. In the foreground boats, blue and white pleasure-steamers with rows of white lifebelts; above them the towpath and trees; then turrets and Tudor chimneystacks and, to the right, Wren's William and Mary front with dressings of white Portland. Above a high wall, the tops of the wych-elms in Queen Mary's Bower and, to the right again, half hidden by willow, the castellated roof of King William's Banqueting House. And, of course, everywhere people: people fishing, people boating, people eating ice cream, people sitting on seats, people lying on grassy banks, people

The Long Water and, below, Bewick's engraving of a fallow deer.

on holiday, people enjoying themselves ....

'The lights were sinking as we paused for a moment upon the terrace that overlooks the river. The steamers were landing their trippers on the bank; there was a distant cheering, the sound of singing, as if people waved their hats and joined in some last song.'

Virginia Woolf: *The Waves*

Part 3 Natural History

Wild Flowers and Trees. Both parks (Hampton Court and Bushy) consist predominantly of dry and acid grassland and support a flora characteristic of such a habitat. Among the more interesting plants are the Upright Chickweed and the Slender Cudweed, which occur in both parks, and the Subterranean Clover, found at Hampton Court, where the Thyme-leaved Speedwell also occurs.

Owing to the nearness of the Thames, and other factors, Hampton Court has more luxuriant vegetation. Its pools are deeper and less contaminated. *Ranunculus circinatus* grows in Long Water. Bushy Park is better provided with rushy hollows, favoured by such plants as Marsh Pennywort. The White and the Fringed Water Lily grow in both parks. Something has already been said about trees. Two more interesting native trees planted in Bushy Park are the Small-leaved Lime and the Crack Willow (*salix fragilis*). In addition to the specimen trees and rarities, all the trees and shrubs commonly found in English parklands grow here. Elms often reach an imposing height. Whenever a tree falls it is replaced.

Deer etc. The fallow deer – some four hundred and fifty of them in the Home Park, are believed to be descended from those in this area in the reign of Henry VIII. The red deer in both parks (about a hundred in all) have been introduced since the last war. They are free to range in their

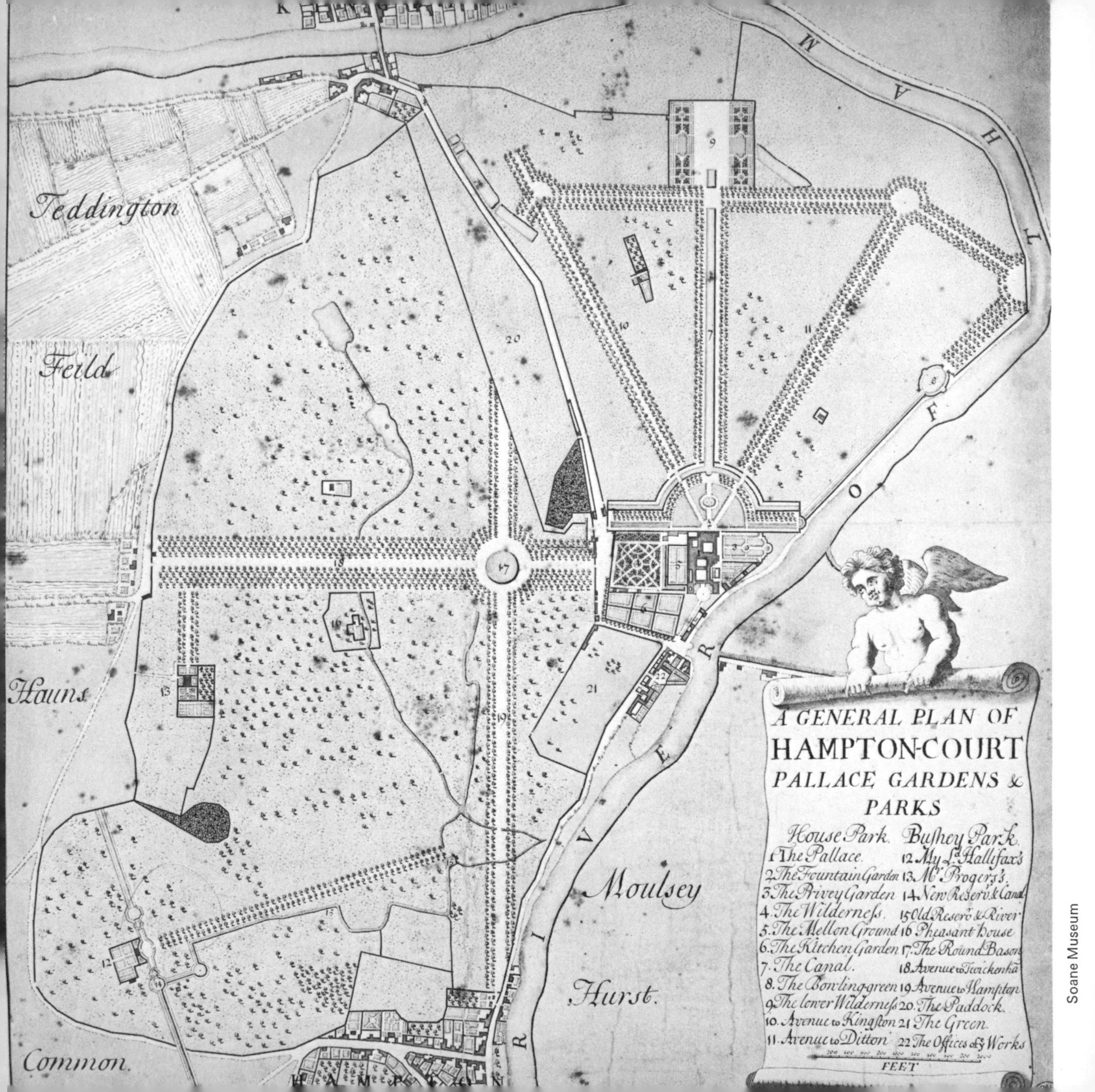
A GENERAL PLAN OF
HAMPTON-COURT
PALLACE GARDENS &
PARKS
House Park.
1. The Pallace.
2. The Fountain Garden
3. The Privey Garden
4. The Wilderness.
5. The Mellon Ground
6. The Kitchen Garden
7. The Canal.
8. The Bowling-green
9. The lower Wilderness
10. Avenue to Kingston
11. Avenue to Ditton
Bushey Park.
12. My Ld. Hallifax's
13. Mr. Progers's.
14. New Reserv. & Canal
15. Old Reserv. & River
16. Pheasant house
17. The Round Bason
18. Avenue to Twickenha
19. Avenue to Hampton
20. The Paddock.
21. The Green.
22. The Offices of ye Works
FEET
Teddington
Feild
Hauns
Common.
Moulsey
Hurst.
RIVER OF THAMES
Soane Museum

Bridgeman's map of Hampton Court and Bushy.
Right: Bewick's engravings of woodcock and jay.

natural surroundings. The plantations are fenced against them and young trees in the avenues are boxed in with wood. Oats, beans and roots are grown for the deer, in Bushy Park. Foxes are occasionally found in both parks. Rabbits persist. Hedgehogs and moles are numerous.

Birds. At Hampton Court and in Bushy Park as many as eighty-two species of bird have been recorded in one year. Among the less common nesters, the Grey Wagtail in Bushy Park has been outstanding; but nests have been found, too, of the Long-tailed Tit, Goldcrest, Tree Sparrow, Swift, Swallow, Stock Dove, Kestrel, Little Owl and Tawny Owl.

A Common Sandpiper was seen by the Longford River in May, 1968. Woodcock were noted on six occasions. Partridges and Pheasants, in small numbers, continue to breed in Bushy Park. One or two pairs of Little Grebe are usually to be found in Canal Plantation. Great and Lesser Spotted Woodpeckers occasionally nest in Bushy Park.

Among the Warblers, the following have been seen or have nested: Sedge Warbler, Garden Warbler, Blackcap, Common Whitethroat, Chiffchaff, Willow Warbler.

In the winter, Redwings are usually plentiful and Bramblings are sometimes seen among the beeches.

Comparatively rare species (near London) have included the Green Sandpiper,

Bewick's wood engravings of the heron and the hooded crow.

Greenshank, Oystercatcher, Hooded Crow, Grasshopper Warbler and Common Redstart.

The commoner tits and finches occur and many of them nest. The buntings are shyer; and although Reed Buntings are believed to have nested in Bushy Park, the Yellowhammer, so common in less built-up districts, must still be considered a rare visitor.

Herons are frequent visitors, as are also Kingfishers. Carrion Crows are too common; but the rookery, which in the limes flanking Long Water had fifty-eight nests in 1946, was found deserted in 1960.

The ornamental waters in both parks support a large population of moorhen and coot, as well as the commoner ducks. Canada Geese are protected and encouraged to nest in the Waterhouse Plantation.

The wood engravings are by Thomas Bewick. They are taken from *1800 woodcuts* (sic) *by Thomas Bewick and his school* (Dover Publications Inc., New York)

Bibliography

Manuscript Sources

The principal sources for contemporary accounts, plans etc. are the Public Record Office, British Museum, Royal Institute of British Architects and the Sir John Soane Museum

Printed Sources

Books

Green, David *Gardener to Queen Anne* (OUP., 1956)

Hamilton, Elizabeth *William's Mary* (Hamish Hamilton, 1972)

Knyff, Leonard and Joannes Kip *Britannia Illustrata* (1709)
Nouveau Théâtre de la Grand Bretagne (1724)

Law, Ernest *The Chestnut Avenue, Bushy Park* (Bell, 1919)
The History of Hampton Court Palace (3 vols., Bell, 1890–1)
Hampton Court Gardens Old and New (Bell, 1926)

Mollet, André *Le Jardin de Plaisir* (1651)

Switzer, Stephen *The Nobleman, Gentleman and Gardener's Recreation* (1715)
Ichnographia Rustica (1718 and 1741–2)

Tijou, Jean *Nouveau Livre de Dessins* (1693)

Brochures and Guidebooks

Bird Life in the Royal Parks, 1971–72 (DOE, 1974)

Hampton Court Palace Guidebook by G. H. Chettle (HMSO, 1974)

The Royal Parks of London by Richard Church (HMSO, 1956)

Trees in Hampton Court and Bushy Park (HMSO, 1963)

Wild Life in the Royal Parks by Eric Simms (HMSO, 1974)

Articles

Bunt, C. G. E. *An Unfulfilled Project for Improving Hampton Court, by Talman* (RIBA Journal, January 1951)

Green, David *Planners of Royal Parks* (Country Life, 1 March, 1956)

Harris, John *The Diana Fountain at Hampton Court* (Burlington Magazine, July 1969)

Lane, Arthur *Daniel Marot* (Connoisseur, March, 1949)

Printed in England for Her Majesty's Stationery Office by W. S. Cowell Ltd, Ipswich.
Dd 504061. K800 9/73.